Working with Public Information Officers

A Supplement to
Explaining Research: How to Reach Key Audiences to Advance Your Work

Dennis Meredith

www.ExplainingResearch.com

Glyphus LLC

Copyright @ 2010 by Dennis Meredith

Published by Glyphus, LLC
4159 Summit Road
Purlear, NC 28665

www.glyphus.com

All rights reserved. No part of this publication may be reproduced, stored in a retrieval system, or transmitted, in any form or by any means, electronic, mechanical, photocopying, recording or otherwise, without the prior permission of Glyphus, LLC.

Meredith, Dennis.
Working with Public Information Officers

ISBN 978-0-9818848-4-4
1. Communication 2. Science 3. Research

To order additional copies visit *www.WorkingWithPIOs.com*

CONTENTS

1	A Sales Rep or a PIO/Journalist?	1
2	PIOs in Different Institutions Face Different Issues	5
	Academic PIOs Enjoy the Most Freedom	5
	Federal Laboratory PIOs Work Within a Bureaucracy	6
	Government Agency PIOs Cope With Politics (Well, Duh!)	7
	Corporate PIOs Work Within a Business Mission	9
3	How PIOs Can Help You	13
	As an Editorial and Media Relations Expert	13
	As an Institutional Ambassador	15
	As an Educator	16
	As a Hard Questioner	17
4	Get to Know Your PIO	19
	Questions to Ask a PIO	20
	Questions *Not* to Ask a PIO	21
5	How to Help Your PIO	23
	Make Hard News Releases Easy	26

	Bring the Feature Story to Life	28
	Understand the Difference Between Media Relations and Public Relations	29
	Take No for an Answer	30
	Advocate for Your PIO	30
	Build Communications into Your Budget	31
	Commend Your PIO	31
	Involve PIOs in Administrative Meetings	31
	Propose Communications Training	32
	Embed Your PIO	32
6	**Work With PIOs Outside Your Institution**	**35**
	PIOs at Journal Publishers and Scientific Societies	35
	PIOs at Your Funding Agency	37
	PIOs with Companies and Commercial Agencies	40
	When PIOs Synergize to Publicize	42
7	**Understand Embargoes, Pro and Con**	**45**
	Make Sure Internal Media Observe the Embargo	47
	Avoid Arbitrary Embargoes	48
	Do Post-Embargo Releases	48

(References available online at *www.WorkingwithPIOs.com*)

1

A Sales Rep or a PIO/Journalist?

Public Information Officers (PIOs) can be invaluable allies in your communication efforts. They can offer expert help and give you access to the communications machinery to reach the media and other important audiences. This guide aims to help you develop the most productive relationships with PIOs—whether they are in your institution, at a journal, in a scientific society, or in your funding agency.

PIOs come in two basic models—the sales rep PIO and the PIO/journalist—although any particular PIO might have characteristics of both. The PIO/journalist is far more effective and credible than the sales rep PIO; so understanding the difference between the two is important because it affects the credibility of your research communication to the media and your other audiences.

A sales rep PIO, like other salesmen, hawks his or her product—your research—with little consideration of its true substance or appropriate audiences. The sales rep PIO usually concentrates more on pitching stories indiscriminately to media and less on thoroughly understanding and explaining your research and targeting communications. A sales rep's news releases tend to be essentially advertisements, peppered with subjective terms like "breakthrough" and "major discovery." As a result, they have less credibility with all your audiences, and not just the media.

The sales rep also believes that the media are by far the most important, if not the only, targets for news releases. However, as is discussed in chapter 2 of *Explaining Research,* a news release has a multitude of uses—as internal communications, as statements of record, as alerts to other researchers, and so on. Sometimes media are really secondary targets of news releases.

You will recognize when a sales rep PIO enters your laboratory when he or she shows only a rudimentary knowledge of your research. The sales rep PIO will talk more about the media pitches for your story than the substance of your work. He or she will talk of "placing" stories, as if salesmanship rather than the quality of the news generates media coverage. You might also hear the thud of name-dropping, when the sales rep talks about top reporters or major media he or she will target.

Sales rep PIOs are notorious for spamming reporters with news releases that do not interest them, says *New York Times* science writer Sandra Blakeslee:

> Science writers are bombarded by information, most of which can be deleted or tossed in the waste basket. The detritus is staggering. Despite all efforts to convince them otherwise, many public information officers still send "news" of promotions, campus "news," badly written press releases or story ideas on subjects completely uninteresting to you. Some are pests. They send regular reminders that they sent you something months ago and wonder if you are "still interested." Some are vaguely shmoozey, just "checking in" to see what kinds of stories you might be looking for.

Such spamming, in fact, reduces the chances that legitimate stories from your institution will be seen by journalists.

The sales rep PIO often fails to realistically assess the significance of research from his or her institution, pitching it relentlessly to media regardless of its importance or even legitimacy.

Tom Siegfried, editor of *Science News* and long-time newspaper journalist, recalls how sales rep PIOs from one unnamed university "notoriously, repeatedly would complain to higher-ups in the newspaper that coverage wasn't what they expected it to be . . . from not having it on the front page when it should have been, to not putting three different titles for a researcher in the story [who has] multiple affiliations." Such harassment is guaranteed to alienate not only science journalists, but their editors.

Sales rep PIOs love the telephone. They especially aggravate reporters by telephoning them to pitch stories or to ask whether the reporter has received a release. Most journalists intensely dislike such calls, unless they are about a truly important, breaking story that the journalist might otherwise miss. They prefer that PIOs selectively email them releases that are reliably significant and well-written, so they know that such releases are worth reading.

So, you know you are working with a sales rep PIO if he or she boasts of extensively phone-pitching your release or widely distributing a routine release on your grant or promotion.

Unfortunately, media-naïve administrators tend to find sales rep PIOs appealing, because of their aggressive, indiscriminate pitching. The administrators perceive sales reps as effective and dynamic, whereas journalists pejoratively dub them "flacks."

In contrast, says UC San Diego PIO Kim McDonald, the PIO/journalist—rather than being a pitchman for the institution—aims

> ... to be more of a resource for the journalists and let them make their own conclusions about the stature of your institution, not shove it down their throats by saying "Look how great we are; we have done this and this." If you act like a reporter within your institution, you understand exactly what is going on and you know the kinds of stories that would interest news media. Then you can become more effective long-term.

A PIO/journalist concentrates on the substance of your research. He or she will show up in your laboratory having studied the background material on your work, formulated cogent questions, and with plans to do a comprehensive story that fulfills all the communications objectives of a news release.

The best PIO/journalists will have a substantive general knowledge of the fields he or she covers. So, while a PIO/journalist might ask you to explain a particular technical point about your work, he or she will know the basics of your field.

However, a PIO/journalist will remain "strategically dumb," says Catherine Foster, formerly media relations manager at the Argonne National Laboratory. That is, the PIO/journalist will ask leading questions aimed at eliciting the most effective lay-level explanation of your work. "If you keep pushing scientists to answer, 'What does that mean?' 'How do you know?' you do get a better story," says Foster.

A PIO/journalist will skillfully manage an interview to use your time efficiently and to get you talking, says Johns Hopkins PIO Joann Rodgers, a veteran journalist:

> I never walk into the company of a scientist like the high school kid who calls up the newspaper reporter and says "I have to do a term paper on X. Can you tell me everything you know about

that?" I always have a purpose, I tell them what I am after, and I have a hypothesis. Scientists like hypotheses. As a journalist, I learned that even if my hypothesis is dead wrong, it gets the conversation started.

Besides knowing your work, a PIO/journalist understands the work of the other researchers in his or her institution and how it fits into the field. Such knowledge gives the PIO/journalist an authority with media that pays off. For example, *San Francisco Chronicle* science editor David Perlman cites UC Berkeley's Robert Sanders as an example of a savvy PIO/journalist: "If I call up Bob and ask who I should talk to about space or any other topic, and if he tells me who I should talk to, I damn well talk to them."

Certainly, a good PIO/journalist will have a touch of the sales rep. He or she will develop a strategy to pitch your story to reporters, but in a realistic, professional way, says Perlman:

> They know what really makes a story for either general reporters or any specific reporter with whom they are engaged in trying to peddle a story... They know the particular reporters and what their interests are. If they do know that there is a major breakthrough for the year, then they need to know how not to oversell it, but to let you know that this a lot more important than you may think.

For example, Perlman cites University of Colorado's Jim Scott as a PIO/journalist who knows how to pitch:

> Maybe a couple of times a month he will call me because they have a press release that is going to be emailed to me, and I might be interested in it. He never once [called] me about a press release that isn't a story.... He is impeccable that way.

What's more, a good PIO/journalist keeps track of the hottest media topics of the moment, says Perlman: "Anybody who has a good yarn that requires the use of the word 'evolution,' I'll fall for it every time. That is because it is a cause now—the woeful problem of teaching evolution in schools with the rise of intelligent design and creationists."

A final hallmark of a PIO/journalist is membership in the National Association of Science Writers or the American Medical Writers Association. To encourage PIO/journalism, those associations publish newsletters and organize conferences that promulgate good communications practices.

2

PIOs in Different Institutions Face Different Issues

All PIO/journalists share the same professionalism. However, they may function differently depending on the type of institution they are in—universities, national laboratories, government agencies, or corporations. Some examples of those differences:

Academic PIOs Enjoy the Most Freedom

PIOs at universities and at academic-type institutions such as NIH typically have more latitude to cover research than do those at the national laboratories or corporations. They can usually choose the people and stories based on their significance and interest to audiences. This basic freedom arises in part because, although university PIOs answer to their vice presidents, those vice presidents ultimately answer to the faculty.

However, university PIOs sometimes face political pressures to concentrate on a particular research program, department, school, or laboratory because the administration decides it needs to be covered. A research program might be at a critical funding juncture. Or an administrator might want coverage of a school or department to mollify a dean or department chair who complains of neglect.

PIOs in medical centers, compared to their university counterparts, operate under more formal restrictions. For one thing, the Health Insurance Portability and Accountability Act (HIPAA) severely restricts public disclosure of information about patients. And in communicating their animal research, medical centers must take into account the threat of destructive actions by animal rights extremists. Thus, medical centers have rigorous policies governing how animal research can be publicized.

Federal Laboratory PIOs Work Within a Bureaucracy

PIOs at the federal laboratories work within a more complex bureaucracy than do those at universities. For example, news releases from federal laboratories must usually be approved up a chain of command all the way to the national-level administration. Such approvals can sometimes be complex. Catherine Foster, recalls the approval situation at Argonne National Laboratory, where she was media relations manager: "Argonne is under the DOE Office of Science, but if we had a research project funded by the Office of Energy Efficiency & Renewable Energy, also part of the DOE, we had to get approval from both offices for a release."

An adept federal laboratory PIO/journalist can navigate such political mazes, in order to do justice to good research. So, if your PIO is sympathetic to your cause, strategize with him or her about the best way to work the system to communicate your research.

Bureaucratic interference with a government PIO's function can do more damage than simply complicating news release approval. For example, National Public Radio science reporter Joe Palca recalls when the Reagan administration instituted a stunningly restrictive rule on press contacts:

"In the mid-80s, the Reagan administration had put in a policy that said no NIH scientist could talk to the national media without approval, not from NIH but from HHS headquarters," recalls Palca. NIH PIO Ann Thomas 'was apoplectic, because in the world we work in, you pick up the phone and you talk with the researcher,' he says. Thomas, considered a consummate professional, did the best she could to lessen the rule's impact.

"She admitted it was a completely indefensible policy and they would pass through requests as quickly as possible. She said 'This is the way things are right now, and we are frustrated. This is a bad time for us but this is the policy.'" That candor and display of journalistic understanding ultimately worked in

NIH's favor, says Palca. "We said, 'God bless you,' and in my opinion she won all the credibility she needed at that moment."

In some cases, federal laboratory PIOs will take significant risks to buck such interference. Foster recalls when a DOE press secretary instituted a rule requiring media calls to any DOE laboratory to be cleared through her, a major headache considering the number of laboratories and the volume of calls:

> We all ignored it, and she didn't like it, but we took the mindset that we will be here after you are gone. Our relationships with the reporters with whom we work with are more important to us than whatever power you have right now. That was a very risky attitude to take but it was one we had to take and fortunately we took it unanimously.

Government Agency PIOs Cope With Politics (Well, Duh!)

PIOs at many government agencies in Washington, D.C. work within an even more restrictive communications environment. They are often required to clear all media requests for interviews up a sometimes torturous and politicized chain of command. They typically sit in on media interviews—an unusual practice at universities. And they work under policies that can change with each change in leadership.

Leah Young is a veteran journalist and government PIO, whose career included extensive service at the Substance Abuse & Mental Health Services Administration (SAMSHA). She recalls the political complexities of government agencies: "A PIO or a scientist coming into an agency really has to know what he is going to be allowed to say—whether he is going to be able to speak his mind within normal limits, or whether he will be tightly controlled."

Government agencies may be staffed with politically-appointed PIOs who have little or no journalistic experience, says Young. Thus, scientists in such agencies should develop their own media relations skills.

Political-appointee PIOs may unfortunately promulgate a cult of personality around the director who appointed them—sometimes to the detriment of the agency. For example, the PIO may try to sell journalists on interviewing the agency director on a story, even when researchers are the real experts on the topic. Young recalls one such instance, when SAMSHA was working with HBO on a series on drug abuse: "My boss insisted that they were only going to talk to the administrator, not the head of the Center for Substance Abuse Treatment, who was an M.D., J.D., M.P.H., and really knew his stuff on treatment."

However, because the director did not contribute any significant insights, his interview was cut from the documentary, which instead featured experts from other federal drug abuse research and treatment agencies. "So we totally missed the opportunity, because there was this determination that they should only interview the administrator," says Young.

Such political interference with media relations can mean more than lost media opportunities. It can also block the flow of critical information to the public. Christopher Jensen of the *New York Times* reported an egregious example of such interference at the National Highway Traffic Safety Administration in 2007. NHTSA administrator Nicole R. Nason, a political appointee, decreed that without special permission NHTSA officials were forbidden to provide information to reporters, except on a background basis. As a result of this stricture, Jensen was not allowed to talk to an NHTSA safety official on the record. In fact, Jensen was told that, instead, he could interview only Nason on the record.

"I declined, failing to see how her appointment as administrator—she was trained as a lawyer—made her an expert in that subject," wrote Jensen. "The agency's new policy effectively means that some of the world's top safety researchers are no longer allowed to talk to reporters or to be freely quoted about automotive safety issues that affect pretty much everybody."

Another notorious example of such political interference was the efforts of NASA's public affairs office under the George W. Bush administration to distort and suppress scientific information on global warming. An investigation by NASA's inspector general's office found that the public affairs office suppressed news releases on global warming and limited the media access of prominent climatologist James Hansen. The NASA inspector general's office's report concluded that "Our investigation found that during the fall of 2004 through early 2006, the NASA Headquarters Office of Public Affairs managed the topic of climate change in a manner that reduced, marginalized, or mischaracterized climate change science made available to the general public through those particular media . . . ."

Despite such political roadblocks, says Young, PIO/journalists at government agencies are nevertheless motivated to help the media, both for the good of their own agency and for the good of the public:

> It is important, especially for agencies that are not well known, to get the name of your organization out, to give taxpayers a reason to want to fund you. Also, many times when I was at SAMSHA, the reporter calling me would be the crime reporter, and not the health reporter. To me, it was extremely important for that reporter to understand

that substance abuse in itself is not criminal behavior; it is a disease. And it is a mission at SAMSHA to get that across, because preventing the spread of substance abuse can be most effective if it is treated as a public health issue. And to get that across, the people who are able to speak to the press have to be the experts.

Corporate PIOs Work Within a Business Mission

Ideally, PIOs at corporations are deeply integrated into the business goals of the company, says Seema Kumar, vice president of global R&D communications at Johnson & Johnson Pharmaceutical Research and Development.

"What I find truly fascinating about the corporate environment is how communications objectives are driven by the strategic objectives of the organization," she says of corporate communications. Such an orientation profoundly affects how PIOs communicate: "Science communications within the pharmaceutical industry is more focused and lower volume than the science communications that is done in academe," says Kumar, whose experience has included work at newspapers, universities, government, and research institutes. In contrast to academe or government, where all published research results and advances are communicated, says Kumar, science communications in corporations focuses on research that is translational, and aligned with the corporate strategy and goals.

"For example, in the early discovery phase of drug research, more can be openly discussed," she says. "But as the drug progresses through early and late development, we have to be careful with how much information we can disclose due to business considerations, such as the protection of intellectual property and competitive information, and regulatory restrictions on claims we can make, or market disclosure requirements. Also, by nature, research and development pipelines are capricious—compounds fall in and out of the pipeline, and you are accountable for any expectations you set about a compound's promise, however early. You don't want to over-promise and under-deliver."

Additionally, "the pharmaceutical industry is highly regulated, and we have to stay within certain parameters. Although in academia, we could speculate and predict what a basic science discovery or a new finding on a compound could mean for a particular disease, we cannot make unsubstantiated claims in the corporate environment, and we have to calibrate anything new we say against past information we have shared with reporters, analysts, investors, and the public."

Even with such strategic restrictions, an effective corporate PIO/journalist knows how to highlight the company's research, says Kumar: "In R&D communications, we are less likely to tell stories about the 'gene of the week,' and more likely to tell stories about trends. We may tell stories about whether personalized medicine is hype or hope; about the changing landscape of clinical trials; or about scientific innovation taking place in emerging regions."

Corporate PIOs provide their researchers more substantial communications guidance and support than is usual in universities or federal laboratories, says Kumar. For example, she says, it is not enough for a corporate researcher to know just how to explain complex science in understandable terms or how to handle tough questions. "That is necessary, but not sufficient," says Kumar. "Corporate researchers must also know how to manage the sensitive questions about related topics such as pricing, cost, reimbursement, and policy, for example." And, to ensure the researchers are prepared, corporate PIOs provide more substantial media training and speaker training, to help researchers cope with the controversies they are likely to encounter, especially in the pharmaceutical industry.

"The pharmaceutical industry has faced many challenges over the years; the science is often very complex, the timelines are long, and there are multiple regulatory restrictions about what is communicated when and how." Thus, says Kumar, corporate media training "has to do with ensuring that what a researcher says about his or her work is characterized correctly, that statements are appropriately balanced, that he or she is able to talk around regulatory and legal restrictions without sounding evasive, and that the researcher is prepared for questions about the benefit-risk ratio, pricing, access, or the safety and side effects of compounds."

Importantly, says Kumar, corporate PIOs help researchers with executive coaching on internal as well as external communications: "Carefully considered internal communication can enhance employee morale, maintain motivation among researchers, and promote the internal transparency that is critical to problem solving. When you have an organization of 8,000 people, you need to ensure that all employees feel committed to the overall mission and vision, and that they are all moving in the same direction." This is particularly important in the R&D world, says Kumar, because scientists typically "are not the type to 'toe the party line' and you need not only to win their hearts but their minds. You have to articulate a compelling strategy for what we want to accomplish, and let the scientists figure out the best scientific way to meet that strategy."

She says that ensuring that there is an environment that encourages scientific creativity and innovation, while channeling that creativity to an overall grand challenge—like curing cancer, or AIDS—that every scientist can commit to is important. "This is where internal communications comes into play, and it can be powerful."

Corporate PIOs also must work within a more stringent legal and regulatory framework than do academic and government PIOs. They must vet all communications through legal and regulatory review processes and balance the requirement to address issues of intellectual property and federal regulations with the need to produce a journalistic-quality news release.

Nevertheless, emphasizes *Science News* editor Tom Siegfried, experienced corporate PIO/journalists can be just as valuable to reporters as are those in universities or federal laboratories: "They are people who know the field; they know the journalists; they know what is helpful," he says. However, Siegfried warns against the corporate practice of short-term jobs for PIOs:

> Some corporations view the public information job as kind of a rotating job that they move people through as a part of their training. And it is not an illegitimate management technique. You want your upper executives to be people who have had different roles in the company so they can understand the company.
>
> But if the PR person with the corporation you are dealing with is there for a year-and-a-half, and then they move somebody else into the job, you have to start from zero developing that relationship of trust and their knowing what your needs are. Corporations that have PIOs who have been there long-term enable you to develop a relationship in which you always know you can get needed information.

3

How PIOs Can Help You

While PIOs in different institutions may operate in different organizational environments, they can all offer the same kinds of services: as an editorial and media relations expert, an institutional ambassador, as a teacher, and as a hard questioner.

As an Editorial and Media Relations Expert

The editorial and media relations services they can offer include

- **Writing and distributing news releases.** PIOs usually write and distribute the many kinds of news releases discussed in chapter 9 of *Explaining Research*: the hard news release, the feature release, the backgrounder, the personal profile, the Q&A, the news tip, the media alert, the grant/gift announcement, and the award announcement. They also distribute them to external media as well as in-house newspapers, newsletters, and institutional Web sites.
- **Pitching feature story ideas.** Your PIO can also develop and pitch exclusive feature stories on your work to particular journalists. The PIO will

likely know journalists' interest and the policies and etiquette of pitching exclusive feature stories.
- **Photography, multimedia, and social media services.** Some news offices have built-in photographic and media services, while others work through separate offices. In either case, your PIO can usually help coordinate photography, video, graphics, and animation support for your communications. PIOs may also help you with Web media relations, including Web video and such social media as blogs, Facebook, and Twitter. For example, Ohio State research communications director Earle Holland helps researchers set up Web cams for media interviews. "That way, when a reporter calls up, they can have the virtual equivalent of a face-to-face interview," he says.
- **Clipping service.** Your news office can provide media clippings resulting from news releases on your work—sometimes a paper clipping, but more often URLs of online stories. The office likely subscribes to a clipping service, such as BurrellesLuce or Cision. Some services also provide video clips and transcripts of television news segments. Although these services offer the most comprehensive source of clippings, you can compile considerable media coverage yourself by subscribing to the free Google News Alerts.
- **Media strategy.** Many PIOs help researchers develop broader media strategies, beyond doing news releases. They can offer advice on persuading reporters to do feature stories on your work and on developing visuals and multimedia. They can also advise you on strategies for handling particular stories and working with reporters.
- **Journalist briefing.** A prepared journalist is much easier to work with than one who walks into your laboratory cold. Your PIO can help with that preparation, especially for complex or controversial stories. For example, research communicator Cathy Yarbrough recalls the preparation she offered reporters before interviews with Francis Collins, head of the Human Genome Project. "Because his time was so precious, if there was a reporter who really wasn't up to speed on the Human Genome Project, I would spend up to an hour on the phone prepping them and giving them the facts so that Francis did not have to recite some of the basics," she says.
- **Journalist scouting.** Your PIO can also brief you on the expertise and target audience of particular reporters. Says communications consultant Lynne Friedmann, as a PIO, once she scouted the journalist's level of understanding, "I can then go to the scientist and say, 'This reporter has never written about this topic. Here is the level at which you need

to present things.' Or, I can say 'This person has read every paper you have ever written, their grasp and understanding of this issue is on a post doc level; however, you need to recognize here is the audience they are writing for.'"
- **Handholder/counselor.** More broadly, your PIO can act as a communication counselor giving you confidence in your ability to work with reporters. "The PIO should be, and often is, a really important bridge between the reporters who are covering science and the scientist who may be scared to death of reporters," says *San Francisco Chronicle* science writer David Perlman.
- **Media credibility.** A reputable PIO, as an institutional spokesperson, also gives you credibility with reporters. As journalist/author Keay Davidson says, "I tend to be suspicious when a scientist contacts me directly and says 'I am working on this interesting research, and I think you should write about it.' So I would urge them to work with their publicist because I am more trusting when news comes from the institution rather than the individual."
- **Crisis communications management.** Your PIO is likely a member of a communications team that should be among the first contact you make when you encounter a crisis. As covered in chapter 25 of *Explaining Research*, managing a crisis—whether an accusation of research fraud or a laboratory accident—means understanding its scientific, legal, ethical, and political implications. Your institution's communications team likely has extensive experience in handling these complexities.

As an Institutional Ambassador

Besides offering media relations services, your PIO can be a useful institutional ambassador because he or she has frequent contact with the administrators whom you want to appreciate your work. For one thing, the PIO can advertise the importance of your research, in a way you might not feel comfortable doing.

"Scientists sometimes feel awkward saying 'I want to tell you about my cool new research,' but as communicators we can do the bells and whistles and jump up and down in proclaiming their work," says Cathy Yarbrough.

Your PIO can also convey to the administration the positions researchers are taking on important issues. "There is often a big chasm that divides university leadership and the faculty," says Ohio State's Earle Holland. "We can represent that leadership before the faculty. We listen to the faculty in ways that they don't get listened to a lot of times. They have a familiarity and a comfort level with us. They will tell us what aggravates them; what they think is wrong. We can take that information back to our leadership and tell them what faculty are feeling."

For example, University of Wisconsin PIO Terry Devitt recalls a message he conveyed to the administration that significantly aided the university's research: "At one point, we weren't devoting enough resources to animal facilities, and we were losing critical people. We were at risk of losing our accreditation, because things weren't being done as they should be. Somebody needed to take that message to the top, and who better than the PIO, who would have to clean up the mess if it were to become public."

PIOs can also link you to offices that can materially advance your research. Vanderbilt PIO David Salisbury recalls one such instance: a graduate student in the laboratory of chemist Sandra Rosenthal had created infinitesimal quantum dots that could be used to produce white light from LEDs—a holy grail of lighting technology. Rosenthal, who had been using the dots as biological markers, had not considered a lighting application until Salisbury did some digging—including contacting a fellow PIO at a national laboratory. He decided to do a news release emphasizing a lighting application.

"Before we began working on the release, Sandy had sent a patent disclosure to our tech transfer people, but they panned it, saying they didn't see anything there," says Salisbury. "But when the release got broadly picked up and the researcher began receiving emails and phone calls from a number of companies, they suddenly got interested," he recalls. The advance ended up getting a *Popular Mechanics* 2006 Breakthrough Award.

"After the coverage died down, Sandy freely acknowledged that if it wasn't for our news and media relations all that never would have come about," says Salisbury. "It's sometimes the case that many researchers may be too close to their work to see broader applications that others, like a PIO, can, see," says Salisbury. "In this case, the lighting industry was just outside their frame of reference."

As an Educator

A PIO/journalist can also be an effective communication educator, for example showing how to develop clear research explanations, says Holland: "We teach by example," he says. "When we get the response from a faculty member on a draft news release, they sometimes say 'You put it together in a way I wouldn't have done it, but it makes it so much clearer to do it that way.' That can open a conversation about, for example, why we might have chosen to leave out certain caveats that weren't needed."

More generally, PIOs can also teach about the nature of media, says Johns Hopkins' Joann Rodgers. For example, she tells the PIOs under her to use faculty complaints about inaccurate media coverage as "teachable moments": "I tell them to ask the faculty member what happened," she says. "Generally, he will not be able to think of a single instance of inaccurate coverage, or it was a repeat of something he heard. Sometimes, he will really have a story to tell, and the PIO can help him. Maybe the faculty member or the journalist did something wrong. The PIO can help the faculty member figure out how to make sure the problem doesn't happen again, or call the journalist to figure out the problem and fix it."

As a Hard Questioner

Finally, as a hard questioner a good PIO/journalist can confront you with those tough questions that you must answer if you are to preserve your reputation and advance your work.

"We are not stenographers," declares Rodgers. "We have to ask tough questions sometimes that scientists or administrators do not like to answer." Rodgers says such tough questions involve challenging "the wisdom of not telling people what they need to know to make an informed decision, or of any omission that is misleading to the public." Such hard questions can cover inadequate disclaimers about your work, not clarifying conflict of interest, and even that ethically dubious motive behind a breakfast that a Wall Street analyst invited you to with the national reporter, says Rodgers.

"It is about reputation management," she says. "You can't get people's attention if you don't have credibility, and the way you get credibility is to be open."

4

Get to Know Your PIO

Researchers almost never take the initiative to get to know PIOs, even though they can be valuable partners in disseminating their research. As Duke communications director David Jarmul says,

"I frankly find it both disappointing and even a little shocking that I don't get more calls or emails from researchers saying, 'Hey, would you like to just come by sometime and have a cup of coffee, and I can show you what we are doing?' Even if nothing is breaking, I just like to know what researchers are up to."

Getting to know your PIO enables you to understand the strengths and limitations of the people and office that help link you to the public and to fellow scientists through professional media. So, perhaps invite your PIO for a laboratory tour and/or chat over lunch.

Such interaction, says Catherine Foster, will "put a face on the research, help us understand why the work is important, and that way we can be better advocates with reporters and members of the public for what you are doing."

Your initial contacts represent only the first step in a long-term partnership between PIO and researcher, emphasizes Ohio State PIO Earle Holland:

> I don't just care about the first story; I care about the twenty stories I am going to do during my career on that work. Once we establish that there is some substantive work there, then we can talk about

possibilities, depending on what type of research it is, whether the science is good and interesting. Does it have that gee whiz effect? Is it about something like volcanoes or puppies that are "magic" stories? Or, if it is something abstract or abstruse, I don't want to give false expectations.

Questions to Ask a PIO

In getting to know your PIO, here are some useful questions you can ask:

- **What is your level of understanding of my research area?** You can do some scouting to answer this question by checking the news office Web site to see the kinds and level of news releases your PIO has done. Also, look for a bio that indicates the PIO's experience and training level. And when you meet the PIO, ask about his or her understanding directly, perhaps phrasing the question diplomatically as "What information do you need about my work?"
- **Where is your office in the organizational chart?** This hierarchy may well influence whether your PIO is a sales rep or a PIO/journalist. If the office is under a vice president for communications, its culture is more likely to be journalistic. However, if the office is under marketing or development, the PIO may face more pressure to "sell" the institution. Such PIOs may find themselves not only writing news releases but promotional materials for donors. The hype language of such promos may tend to leak into news releases.
- **How is the news office structured?** The office may have a beat system, in which each PIO covers research in specific fields. Knowing what beats your PIO covers might give clues to how significantly the office views your work. For example, if your department is lumped into a grab bag of disparate areas on a generalist writer's beat, that writer is less likely to cover your work adequately.
- **What are the office's news policies?** News policies differ according to the type of institution. However, even the same types of institutions may have significant differences, sometimes even eccentric ones. For example, a major medical center's news policy once had a policy prohibiting revealing in news releases the animal species used in experiments. The releases could only refer mysteriously to a "laboratory model" used in the experiments. That news office was under the vice president for development, who feared alienating donors who opposed animal

research. Other institutional policies might differ on whether journalists can contact researchers directly or must go through the news office; or whether journalists can visit laboratories freely or require an escort.
- **What services does the office provide?** Ask what services the office provides. Learn the procedure for using those services, what communications products you will receive, any costs, and the key contacts. For example, some news services produce only low-resolution Web videos using modest equipment, while others are organized to produce broadcast-quality video news packages.

Questions *Not* to Ask a PIO

There are also questions you should not ask your PIO, unless you enjoy aggravating people. Some examples:

- **Will my story get a lot of media coverage?** A PIO can only guess at the level of coverage of a story. On a slow news day, a run-of-the-mill release might garner the lead slot in a newspaper or Web site. In contrast, for no apparent reason, a dramatic research advance can also disappear down the media maw without a trace. Even the best PIOs admit their shortcomings in prediction. Says the University of Wisconsin's Terry Devitt, "When I try to forecast an outcome, I tend to fail miserably. We all have stories we thought were really good and would go like crazy and others that were kind of dull but that, when they did go out the door, spread like wildfire."
- **Why did article X not mention my research?** This is not a useful question, because any article on a given topic may not include many prominent researchers in the field. Even in-depth feature stories are limited in scope and size. They attempt to give only a brief overview of a topic from a limited viewpoint and thus include only a few researchers. Your PIO almost certainly could not have known about a given article beforehand, because reporters do not advertise their decision to do an article on a given subject beyond their editors. And the reporter's research for that article is limited and even somewhat serendipitous. Unless your PIO contacted every editor of every relevant publication during every publishing cycle, he or she could not know what articles are being planned. And even if the PIO attempted such an outlandish activity, editors would likely not share that information. What's more, the PIO who attempted such an exhaustive and annoying canvas would

immediately be persona non grata in every editorial office. The most productive approach to gaining publicity, which PIO/journalists use, is to assiduously identify the best stories for the media and let editors know about them by producing high-quality news releases and other materials. Your PIO might also maintain contact with journalists by going to scientific meetings, where he or she can find out about reporters' interests and attempt to serve those interests with stories about your research and other work from your institution.

- **Why isn't there a story in the newspaper about my exciting new grant (building, research gadget)?** Money stories do not excite reporters, unless there is embezzlement involved. Nor do bricks and mortar or new research machines—unless the building falls down or the machine explodes. Reporters generally cover only the largest grants, buildings, or research instruments, and those stories are typically perfunctory. As excited as you are about your new grant, building, or gadget, it is not of great interest to the public. So, rather than harassing your PIO to gain news coverage for a grant or building, suggest feature stories on the research funded by such grants or housed in a new building. These stories can reach important audiences by appearing in internal publications and Web sites. And they might even prompt a reporter to do a feature.
- **Somebody told me at a party last night that they hadn't heard about my latest research. Why isn't the word getting out?** Science communicator Rick Borchelt recalls the frustration of such questions when he worked at Oak Ridge National Laboratory: "I had done a lot of really good data analysis on who was receiving our material and what kind of an impact it was having," he says. "We even did focus groups among some of our key stakeholders. Then our director would go to a cocktail party in the community and some wife of a doctor in the community would say 'I never heard of what Oak Ridge is doing.' Immediately he would ask us to completely change our whole communication focus. It's amazing that scientists whose lives are driven by data, even when they are given hard facts about the impact of science communication, prefer to listen to cocktail chatter."

5

How to Help Your PIO

Your relationship with your PIO is a two-way street—a broad highway, one might hope. Your PIO can provide you invaluable services, but he or she also needs help from you. Fortunately, that help is cheap in both your resources and time. Your PIO needs only the information to do the job, a good working relationship, and your advocacy.

First, your PIO's information needs:

- **Access to principals.** If you are a senior scientist, be willing to work directly with your PIO, whether you are asked to review a news release or answer a question about your field. Certainly, junior researchers in your laboratory can give the PIO such information, but recognize that you are the one whom the PIO needs to quote in a news release, rather than a junior researcher.
- **Your communication needs and expectations.** Share your fondest communications dream with your PIO. Understanding that dream will help your PIO formulate a release strategy to realize it. For example, a PIO usually thinks more about reaching lay media, but your communications Valhalla might be a one-paragraph news item in *Chemical & Engineering News*. If so, your PIO can make sure the release gets to the right reporter there.

- **Early communications strategizing.** Begin early to work with your communicators to plan a broad strategy for your research program. "It is so often the case that, even with multi-million-dollar projects, that only at the very end do they think about communications," says Duke's David Jarmul. "They should have been working with trained professionals who understand and can do communications properly on the very first day. There are all kinds of communications—producing animations, or making a radio show, or a series of science education modules, or an outreach program with local legislators—that you probably should have done, but you can't when you wait until the last minute."
- **How you fit into the institutional mission.** Your work might be central to the current mission of your institution . . . or it might not. Either way, if you understand where you fit, you can better help your PIO do his or her job. Duke Pratt Engineering School Communications Director Deborah Hill emphasizes the importance of taking a long view: "If a researcher's work is not part of the whim of the day, we can still develop strategies that will help them expand their careers. Ultimately, their work *will* become the whim of the day, and if we don't help them with our communications tools, we are being shortsighted."
- **Good stories.** PIO/journalists make their living off good stories, just like media journalists. While you might think to call your PIO only when you have a paper in *JAMA, Science,* or *Nature,* you should really contact him or her whenever something happens that excites you. Maybe it is a state-of-the-art telescope imager, a graduate student with a fascinating background, a cool behavioral study you are beginning . . . all can make good story fodder. When in doubt, let your PIO know, because he or she can use just about any story in some way—if not as a news release, then as a feature, or even a tip to a journalist. Also, let your PIO know if you have expertise on a topic in the news and are willing to comment for the media. Your PIO can issue a news tip about your opinion and availability. Finally, tell them stories about your colleagues' work, as well. Both your PIO and your colleagues will appreciate it.
- **Early tips.** Give your PIO plenty of lead time on news releases—ideally telling him or her when a paper has just been accepted. PIOs' second-worst aggravation is receiving a note from a researcher saying "I have a paper coming out tomorrow; can we do a news release?" The *worst* aggravation is a note that begins "I had a paper come out last week . . ." While issuing news releases after publication is still useful, sending out a release before the embargo makes it likelier that daily media will

use the release. PIOs need all the lead time they can get, given the time required to set up and do an interview, produce images and video, write the release, get it approved, and distribute it. Do not worry that your PIO will break an embargo; PIOs observe embargoes religiously. And sending out a release or giving media interviews before the embargo is common practice. Do not depend on the journal to tell your PIO about your paper. If the journal sends such notices at all, they usually send them only a week or so before publication. However, bless their hearts, many journals such as *Science, JAMA, PNAS* and Cell Press journals post information on articles when they have been accepted but not yet scheduled.

- **Involvement beyond news releases.** Ask your PIO for help when you get any request for a lay-language summary of your work—for example, as part of press materials for a meeting. Not to insult your writing ability, but your PIO can probably do a much better job. Or, if you want to write for *American Scientist* or *Scientific American*, ask your PIO for help.
- **Comprehensive information.** Besides providing a copy of your paper, also offer background articles such as *Scientific American* features on your field, and even book references. Provide your PIO a "cheat sheet" lay-level summary of your work. The summary can include analogies you have thought up, to test them on your PIO. Supply any visuals you already have, including video and graphics; but also offer ideas for new images and video. As you would with external journalists, put your work in context, for example discussing clinical applications and how your findings help answer broader questions in your field.
- **Quick feedback.** Also important for a PIO is quick response to his or her request for an interview, draft release edit, media interview request, and so on. Such response is especially critical because the Web has accelerated scientific publishing enormously. For example, says Howard Hughes Medical Institute (HHMI) PIO Jim Keeley, "Ten years ago, journals just did not have a mechanism to get a paper out quickly. But now with their own Web sites, when they hear about a competing paper at another publication, they can post a paper on their site within twenty-four to forty-eight hours and call it an immediate early publication."

Following the above guidelines will get you off to an excellent start with your PIO. Beyond that start, here are tips for forging an effective working relationship:

Make Hard News Releases Easy

"Hard news" releases are those that announce research findings published in a journal or delivered at a meeting. To produce an effective hard news release, your PIO needs a productive interview, an efficient approval process, and a well-targeted distribution. The best preparation for these is to read the relevant chapters in *Explaining Research*—chapter 23 on media interviews, chapters 9 and 10 on news releases, and chapter 11 on news release distribution. Here are some other steps you can take to prepare for your collaboration with your PIO:

- **Find out your PIO's level of understanding of the field and your research.**
- **Provide background accordingly, ranging from URLs of basic articles on your field to technical review articles.**
- **Find out the PIO's level of science writing experience.** While one PIO might have decades of experience in writing about research, another might be fresh out of journalism school with little scientific background. If it is the former, you can assume that the PIO will deliver a usable draft; if the latter, you may expect multiple iterations to get the story right.
- **Ask how your PIO will gather information in the interview—whether recording and transcribing the interview or taking notes.** If the interview is recorded, feel free to rattle off concepts and quotes as quickly as you like. If the PIO takes notes, be more deliberate, so that the notes will be more accurate, as will the draft release.

Providing clear lay-level explanations and analogies during the interview will be particularly important. You may, in fact, have to take the lead in supplying such components, because PIOs are sometimes diffident, says Johns Hopkins' Joann Rodgers:

"Some PIOs have the perception that science is very serious stuff, and therefore explanations have to be complex, difficult, arcane," she says. "They often self-censor because their perception is that they have to be as eggheaded as the scientist, or else they won't be respected."

Veteran PIOs also can sometimes be too immersed in a particular field, and lose track of the need to keep explanations at a lay level. Research communicator Cathy Yarbrough, while working at the American Heart Association, had this "inside-baseball" problem with long-time staffers.

"One of the biggest challenges I had was with communications staff who knew 'too much.' Especially those who had a lot of experience on the job; they didn't understand what lay people didn't know," she says. "They would write releases that referred to 'cardiac events,' without defining what a cardiac event was. I would joke with them: 'What is a cardiac event? Is it falling in love? Is it the heart association charity ball? The Heart Association marathon race?'"

After the interview, the PIO will write a draft release for your approval. Some tips on making that process effective:

- **Do not suffer a rough draft.** Even a draft release should be tightly written, with its punctuation, spelling, and grammar checked before you get it. However, sometimes a novice or less-than-professional PIO will try to use the researcher as an editor-of-*first*-resort. Typically, such a PIO will tell you something like "Here's a rough draft that we can go back and forth on." Return such rough drafts, requesting a clean-up. You are actually doing the PIO a favor. Requiring a polished draft helps the PIO learn good editorial practices, and it also saves your colleagues from the aggravation of rough drafts.
- **Provide a single point of contact.** The PIO should not have to combine and reconcile multiple, conflicting drafts from different researchers. So, produce a consensus version before sending the release back to your PIO.
- **Respect the news release style.** As discussed in chapter 10 of *Explaining Research*, a news release needs to be an accessible lay-level explanation of a piece of work and cannot contain all the technical details or caveats of a scientific paper. If you feel such detail really needs to be part of the public record, prepare a more technical release for your Web site. If you find yourself doing a wholesale rewrite of a perfectly well-written release, consider whether you are making only stylistic changes, rather than truly factual corrections. If the former is the case, perhaps you are reacting to the lay-level tone, rather than to the substantive facts.
- **Do not accept hype or overselling.** Even though the release needs to be in an accessible, lay-level style, that does not mean engaging in hyperbole. Do not feel you must approve a release that has hype or overselling that makes you uncomfortable.
- **Negotiate editorially as equals.** Treat your PIO as you would any other professional, respecting his or her position on differences of opinion. As Leah Young says, "Very often I found in writing press releases that the scientist wants this to be very scientifically accurate to a point that he wants to use terminology that nobody but another scientist in his field will understand. A good PIO can work with the scientist, saying 'If you don't like the words that I came up with in translating what you are doing, let's see if together we can come up with words that satisfy both of us.'"

- **Do not freak out over details.** Mistakes can be fixed, even egregious ones. "I have had scientists who have seemed to think something is unfixable, when it is really a matter of a single word choice or a different phrase," says Duke research communicator Joanna Downer. "A scientist once called me declaring adamantly that a release was 'just wrong.' But when we went through it, there was one inaccuracy, which was fixed by a different phrase. And he had gotten so worked up about it, he just couldn't see what was wrong and what was right."
- **Understand "brand name" issues and sensitivities.** There may be political ramifications to how the release cites your institution, department, and title. Tell your PIO about sensitivities you are aware of, and understand the sensitivities the PIO must cope with.
- **Avoid "approval hell."** Try to keep to a minimum the number of people who need to approve a release, although policy may not give you any say in the matter. Universities typically only require approval from the faculty researcher. However, national laboratories and corporations for political and business reasons require much more elaborate approval processes. Such a review need not be onerous. For example, Cathy Yarbrough recalls the efficiency of the review of news release drafts at Novartis when she worked there: "Everybody who had a role in the approval sat at the table at the same time. Regulatory couldn't change something in a way that legal would object to, or medical would object to. And it was so efficient to deal with it this way; also it was a way for me to learn the issues." In contrast, your institution might subject releases to "approval hell," in which multiple administrators decide they need to approve a release only to "scent mark" it, rather than offer useful input. PIOs suffer among the worst torments of approval hell when they find themselves trapped between administrators who disagree on edits, notes Yarbrough.

Bring the Feature Story to Life

A feature story about your work offers an engaging inside look that a news release cannot give. Even if a feature is only for "internal" publications and not for release to external media, it could be widely read by administrators, donors, students, and other important audiences. Also, it will likely be seen by journalists. Just as there is no such thing as an "internal" news release, there is no such thing as an internal publication—especially given the reach of the Web.

You can help an in-house writer craft a good feature by providing human interest stories, "you-are-there" descriptions, conceptual background, and potential applications of your work. So, in your interview offer personal background, anecdotes, opinion, and other information that will bring a feature to life. Offer the writer laboratory tours, field trips, and even participation in experiments—assuming the experiments do not involve organ transplantation or high-voltage electricity. For example, when I gathered information for in-house articles, researchers gave me the opportunity to witness the launch of a large sounding rocket, to helicopter over the giant Arecibo radio dish in Puerto Rico, to have my brain waves measured, and to capture bats in a Costa Rican rain forest. For more detail on the journalistic elements of a good feature story, see chapter 16 of *Explaining Research*.

Understand the Difference Between Media Relations and Public Relations

Knowing the distinction between media relations and public relations can give you a more realistic approach to working with your PIO. *Media relations* pertains to seeking coverage in newspapers, magazines, Web sites, and other outlets. However, *public relations* encompasses a broader spectrum of communications with your audiences, of which media relations is only one component. This distinction is important, because it implies that your communications should involve more than just getting stories in the *New York Times* or any other media outlet.

You might have made a "ewwww!" face when you read the term "public relations," because that term does carry a disreputable connotation. However, you should think of public relations, not as slimy salesmanship, but in its professional sense as a set of strategies for communicating your research and its importance to all your audiences.

While your PIO will be versed in media relations, he or she is not a "public relations" person. Your institution does employ public relations people who oversee the broader range of communications. However, they will not likely be at your disposal for your research communications. You are responsible for your own public relations strategy, and you do need a broader strategy than just one of media relations. Do not fall into the trap that public relations consultant Lynne Friedmann describes in some short-sighted clients:

"They do not understand that media relations is a tactic of public relations; that it is more than publicity. Their perspective is short-term, and they're looking for the big media score." Rather, as is discussed in *Explaining*

Research, you should plan a long-term strategy for communicating your work in a credible way to all interested audiences.

Take No for an Answer

Your PIO will quite often answer "yes" when you ask for a news release or other communication. But the answer may sometimes be "no," and for good reason. For one thing, your story might not be ripe for publicity.

"We will only do a release if the work has been through some level of peer review, made it to a plenary session at a national meeting, published in a journal, or about to be published in a journal," says Don Gibbons, formerly at Harvard Medical School and now chief communications officer at the California Institute for Regenerative Medicine. "Sometimes, though, we get cases in which people want publicity when their work is at a very early stage. For example, surgeons are starting a trial, and they want more referrals. In such cases, publicity is usually not appropriate."

Also, your work—albeit excellent and scientifically significant—might be too technical or narrow for the broader audience that a PIO seeks to reach.

"People may be doing just absolutely magnificent research, but you can't understand it, you can't translate it," says Ohio State's Earle Holland. "Even research in molecular biology and cancer has gotten to the point that, while it is incredibly important, you have to understand three chapters of a textbook before you can really understand its significance."

In cases where your work is just not appropriate for lay-level communication, you will be responsible for producing the articles, Web content, and other materials for reaching important technical and professional audiences.

Finally, your PIO might say no because of the considerable demands of serving the communications needs of a multitude of researchers. In such cases, your best recourse is to make it as easy as possible for the PIO to say yes—by following the guidelines for effectively working with your PIO discussed in this guide. Also, a yes is more likely if you make the PIO's life easier by developing your own quality presentations, Web site, multimedia, and other content.

Advocate for Your PIO

With only modest effort, you can greatly help your PIO and your communication office do their jobs by advocating for them. Such advocacy not only helps your own work, but raises all the "institutional boats"—advancing the mission of your university, corporation, or government laboratory or agency. Here are some of the things you can do:

Build Communications into Your Budget

When planning your budget—whether for a specific project or your laboratory, department, or center—include money for communications, says Duke's David Jarmul.

"The increasingly multimedia nature of science journalism is a big challenge for universities and scientists," he says, "It puts tremendous pressure on us from a budgetary and personnel standpoint. It costs a lot of money to produce animation, slick multimedia features, and a professional Web site. And the bar is set high, in that everybody wants their Web site to look like *National Geographic*'s.

"There is a reason why large corporations spend a good percentage of their budget on communications and marketing," says Jarmul. "I am not saying that a university should become Proctor and Gamble selling soap, but any communications effort is going to be less effective if it's treated as an afterthought."

Commend Your PIO

If your PIO produces a great news release, feature story, or research animation, drop him or her a note of thanks. Copy that note to appropriate colleagues and administrators. Such kudos will help your PIO get more stories from your colleagues, and perhaps more resources from administrators. The resulting communications enhancement will benefit you, your colleagues, and your institution.

Involve PIOs in Administrative Meetings

Give your communications experts a seat at the management table for any significant decision. All such decisions have a communications component—from building a new laboratory to coping with a research scandal. The PIO can offer not only communications advice, but a useful independent view of the issue. And the experience will educate the PIO.

For example, when the University of Wisconsin planned its stem cell research initiative, the researchers invited PIO Terry Devitt to participate.

"They knew that it was controversial, and in anticipation I sat in on meetings of the bioethics committee," he says. "I could hear them ask the key questions about research on human subjects and conflict of interest."

As a PIO at Novartis, Cathy Yarbrough found that participating in managerial meetings enabled her to communicate more effectively about the company's products.

"Each drug has its own team and as a communications person I was on multiple teams. As such, I learned about the marketing plans, the medical concerns, and other important issues about each drug first-hand," she says.

Propose Communications Training

Communications training teaches you to explain your research clearly not only in media interviews, but in other public venues. By proposing that your group hold such training sessions, you do a favor to yourself, your colleagues, and your PIO. The sessions need not take much time. For example, the Duke News Service hosts periodic half-day training sessions for faculty, covering such topics as how to give a TV interview or write an op-ed. The sessions are conducted by staff PIOs, and they provide faculty useful communications skills. They also introduce researchers to the communicators who can help them with news releases, features, and other communications.

Even simpler, invite your PIO to a laboratory or departmental meeting to talk about the communications process and answer questions. Ask your communicators to participate in new-researcher orientation. For example, at HHMI's orientation for new investigators, vice president for communications and public affairs Avice Meehan briefs the researchers on communications issues. She explains the communications office activities and how researchers can work most effectively with its PIOs to reach important audiences with their research news. Thus, even though HHMI's investigators are spread in universities throughout the country, they are plugged into the institute's communications system from the beginning of their tenure as investigators.

Embed Your PIO

Universities and other institutions have an unfortunate habit of isolating PIOs in administrative buildings or even in off-campus offices. For example, when I worked at Cornell, the news office operated out of a strip mall two miles from campus—next to a real estate office and a Dunkin' Donuts.

Encourage administrators to house your PIO and news office near faculty offices and laboratories. Or, at least provide your PIO an office as a base of operations in your laboratory building. UC San Diego PIO Kim McDonald finds that his office in the natural sciences building greatly aids interaction with researchers:

> I see the scientists in the hall, in the elevator, at lunch, and when I'm walking across campus. I can ask them what they're working on. It is like being a police reporter and working out of the police station. It is a much more effective way for me to get information. When I do need to contact somebody, if they are not answering their phone I can hunt them down. Having me on campus is a constant reminder for them that "Oh yeah, I have to tell Kim about this paper that I just submitted because I think it is interesting." My constant presence reminds them what I am here to do, which is necessary because their focus is to publish, get tenure, teach classes.

Also, consider embedding your PIO in field trips and expeditions, perhaps as an expedition communications officer. Such embedment can have considerable advantages, because a PIO can do a much better job than you of writing compelling descriptions of the experience and producing quality visuals and audio.

For example, in 1999 and 2000, Duke science writer Monte Basgall was embedded in deep-sea expeditions to the Pacific Hess Deep and Mid-Atlantic Ridge. The Pacific Hess Deep is a Grand-Canyon-sized underwater chasm; and the Mid-Atlantic Ridge is one of the world's largest undersea mountain ranges.

On those expeditions, Basgall chronicled the explorations in real time for their Web sites and wrote news releases and magazine articles on the expeditions and their findings. In fact, he helped the Hess Deep expedition make a major news splash (pun intended), when the researchers discovered a "Lost City" of gigantic undersea spires formed from the minerals spewed from hydrothermal vents. Basgall wrote a news release aboard the ship that NSF and the expedition issued, and the release gained international coverage.

As another example, while at Michigan State University PIO Sue Nichols took part in expeditions to China and Rwanda. On the China expedition, she and a videographer accompanied researcher Jianguo Liu on a trip to the panda habitat in the Wolong Nature Reserve. She recounted the expedition's communications payoff in an article in the National Association of Science Writers *ScienceWriters* newsletter:

> The results: seven online dispatches from the field that we also featured on NSF's home page, some 900 photo images, 10 hours of

professional video, and a Web site accessible to a wide audience that carried the message that science is fascinating, fun, and compelling. Both MSU and NSF still extensively use the materials, as have schools, journalists, and others.

Nichols journeyed to Rwanda to cover the university's effort to develop a specialty coffee crop in the country, to enhance its economy. She describes the impact of her efforts:

> Not only was I able to document a remarkable story, I could show, in a highly personal way, the incredible commitment of our university's faculty, and the amazing contribution such a commitment can make.
>
> ... we wrote up an expansive plan that cast my department as a driver to illustrate MSU's land grant role, its global impact, and to sell coffee. I not only wrote stories for the Web, but also for grant applications, awards, and partners. I wrote coffee labels, brochures, tent cards, and radio promos. My photos appeared in major newspapers.
>
> MSU alone has sold more than a ton of Rwanda specialty coffee, contributing a portion of each sale directly back to the project.

You can embed your PIO in any significant scientific event, and not just a major expedition. For example, in 2003, I was embedded in a week-long "Rubber Boot Camp" at the Organization for Tropical Studies' (OTS) La Selva biological station in Costa Rica. The OTS hosted the educational event—which gave participants a chance to try their hand at field science—to commemorate its 40th anniversary.

As an embedded PIO, I covered the event, writing a feature story about the OTS for Duke's alumni magazine. Also, embedded photographer Chris Hildreth produced "Postcards from Costa Rica," photo-caption gallery depicting OTS research.

6

Work With PIOs Outside Your Institution

Besides your own PIO, you can work with PIOs at journal publishers, scientific societies, and your funding agency to communicate your work. Also, your institution may use commercial PR agencies. Each of these types has differences, and understanding them can help you best benefit from their expertise.

PIOs at Journal Publishers and Scientific Societies

Many journal publishers and scientific societies have media relations offices that publicize the journal's articles or the society's meetings.

PIOs for journals typically send out a media package for each issue containing summaries of papers that might be of interest to journalists. For example, the news operations of *Science, Nature,* and the Cell Press and the American Chemical Society and American Physical Society journals all issue summaries of selected papers they deem newsworthy.

If you have a paper in press, check with your editor about the publicity process for that journal. And, check the journal's Web site for instructions for authors on how best to work with them. For example, American Physical

Society's Web site includes instructions for authors on publicity and outreach. The APS also operates the online magazine Physics, which features lay-level summaries, written by scientists, of important papers in its journals. And *Nature* posts a Web page that explains communication and media benefits.

And, as discussed previously, always *always* **always** let your own PIO know when you have a paper accepted.

Most major journals will notify your PIO when a paper is scheduled for publication. Also as discussed previously, *Science, JAMA, PNAS* and Cell Press journals (bless their hearts a second time) post information on articles when they have been accepted but not yet scheduled. For example, AAAS notifies some 300 PIOs a week about upcoming papers in *Science*, even phoning PIOs with information on particularly newsworthy papers.

These journals recognize that your PIO is an invaluable ally, since he or she can do a news release that publicizes the journal's papers. There may be some limitations on such notifications, however, says AAAS public programs director Ginger Pinholster: "We are constrained until editorial tells us that the paper is accepted and the publication date," she says. "We can't launch into high gear with our notification while the paper is still in flux. However, it doesn't hurt to call, because you can get an idea of whether it looks like the paper is going to survive peer review and approximately when it might be published."

You can aid the journal's publicity process by offering to prepare a lay-level summary of the paper and to provide photos, graphics, video, or animations that illustrate the findings. Also, offer to check for accuracy any press summary on your paper.

You should make sure you are given a chance to edit any press release about your work or risk being at the mercy of their writers, warns the *New York Times* science writer Cornelia Dean, author of the book *Am I Making Myself Clear?*: "The journals are commercial operations. They tout their paper as a way of promoting the journal. And many times scientists have told me 'I saw the press release, and I just about died.'"

If the journal in which your paper is being published lacks such publicity support, find out the publication date from your editor, and make sure your PIO knows about it.

Do not assume that a journal PIO is aware of the newsworthiness of your paper, especially at publishers with many journals—such as *Nature*, Cell Press, the ACS or the AGU.

"We publish something like sixteen journals that total some fifty thousand pages a year of peer reviewed research," says former AGU communications officer Harvey Leifert. "There is no way we can tell what all the good stuff is.

Titles and even abstracts often don't tell that a paper is newsworthy." What's more, says Leifert, journal editors might not even inform the PIO of newsworthy papers: "We try hard, but with only modest success, to get the editors to tip us to exceptionally interesting papers that they have just accepted."

Thus, you and your PIO can alert the journal's PIO to your newsworthy paper and offer to help publicize it. If your institution does not have a PIO who covers research—such as a smaller college—you can contact the journal PIO for help.

PIOs at scientific societies also publicize their meetings. For their national meetings, AAAS, ACS, AGU, and other major societies produce substantial online press kits and virtual press rooms that feature paper summaries, news conference schedules, and other information. To get an idea of the materials offered at such meetings, explore the EurekAlert! meetings announcements site.

Your PIO may prepare news releases on papers being delivered by researchers at such meetings, providing them for the virtual press rooms. Many PIOs also travel to such meetings, to contact reporters and talk about their institution's research. So, besides notifying your PIO about accepted papers, let him or her know about papers accepted for meetings.

PIOs at Your Funding Agency

The federal funding agencies, most notably NSF and NIH, operate active public affairs offices whose PIOs want to hear about your work. Their Web sites are excellent venues for your news releases and other content, because they carry the unique cachet of a major federal agency.

In communicating the advances in projects they have funded, funding agencies aim to demonstrate to Congress and the public their value, says NSF science communicator Leslie Fink, of NSF's Office of Legislative and Public Affairs (OLPA): "I think of it as a general responsibility of a federal agency to inform the American public—in ways that make sense to them and in context—what they are getting for the money they are spending," she says.

NSF and NIH consider such broad communications so important that their grant-making rules stipulate that the research projects must incorporate efforts to broadly disseminate their results. For example, in its merit review criteria, NSF offers these examples of activities for disseminating results beyond scientific publication:

- Partner with museums, nature centers, science centers, and similar institutions to develop exhibits in science, math, and engineering.

- Involve the public or industry, where possible, in research and education activities.
- Give science and engineering presentations to the broader community (e.g., at museums and libraries, on radio shows, and in other such venues.).
- Publish in diverse media (e.g., non-technical literature, and Websites, CD-ROMs, press kits) to reach broad audiences.
- Present research and education results in formats useful to policy-makers, members of Congress, industry, and broad audiences.

However, says OLPA Director Jeff Nesbit, researchers would do well to think beyond this list when proposing their outreach activities:

> Most researchers choose things that they know have worked in the past, and that they have heard other researchers talk about. So, they tend to focus on broadening diversity in education, or educational materials that might show up in classrooms. The review committees are now starting to look for the more innovative and creative ways to broaden the reach of your research, and one of the easiest ways is to look at mass communications vehicles and podcasts and video and projects of that sort.

NSF's public affairs office is eager to partner with researchers and their PIOs to communicate their work, say Nesbit and Fink. Given that the NSF Web site receives about a million hits a month, the chance to have a news release or feature posted on the site represents a golden opportunity.

However, points out Nesbit, the reach of NSF media efforts extend far beyond the Web site. OLPA can produce releases and videos that are fed directly to media partners' cable channels, Web news portals, and Web sites. Also, NSF may produce media-briefing Webcasts on discoveries involving NSF-sponsored research. NSF's outreach efforts are not limited to the material produced by OLPA, he says: "We promote university press releases just as much as we do our own," he says. "It is not about building up NSF's name," says Nesbit. "I don't want to say we are not interested in that, but it is not our principal mission. Our principal mission is using the NSF's name and platform to try to empower and facilitate the research that is going on out there." To help your release reach media, NSF may post it in a media section of the NSF Web page on EurekAlert!, as well as being featured on the NSF Web site, he says.

To have your research news featured by NSF, your PIO should submit it to NSF through the appropriate staff member, as listed on the OLPA Web site. That staffer may do a version for the NSF site that will include a link to your release. Also, NSF might highlight your release in the "News from the Field" section, with a direct link to the release on your site. The NSF site also seeks photos, illustrations, videos, and audio, says Nesbit and Fink. These will be featured on the NSF site and distributed via its Science360 news service.

You can help NSF reach another important audience, Congress, by producing a well-written annual research summary for your program officer. NSF excerpts these summaries in the report it sends to Congress, as part of the budget process. Unfortunately, many summaries are poorly written, says Fink: "The vast majority of the time they are incomprehensible," she says. "We have to figure out what they were talking about and rewrite them in a way that a typical congressperson can not only understand, but get why the work is important. We realize that researchers have a lot of demands on their time, but they should realize that the science budget is not as protected as it once was, and this is one way they can help protect it." OLPA considers these summaries so crucial to NSF's legislative communication mission that it is rewriting them to make them more understandable and posting them on a special NSF Web page for Congressional staff that lists research projects by state.

The NIH communications offices also work with institutional PIOs to publicize NIH-sponsored research. However, the NIH is more decentralized than NSF, with 27 institutes and centers, each of which has its own communications office. These individual communications offices may have different practices for working with you and your PIO. Check with your program officer and/or the PIOs in your institute to explore the best way to work with its communications office.

The main NIH Web site at *NIH.gov* offers news releases about work funded by all of the institutes and centers. Also, it offers the online magazine "Research Matters," which covers all of NIH's research. Those materials are produced by the Office of Communications and Public Liaison.

The National Institute for General Medical Sciences (NIGMS) exemplifies the benefits of working with NIH PIOs. For one thing, its Web site at *NIGMS.NIH.gov* links to universities' own releases about NIGMS-sponsored research. The NIGMS communications staff identifies those releases by searching EurekAlert! So, such releases should be posted on EurekAlert! and should specifically cite NIGMS as a funding source. What's more, NIGMS features selected releases on its e-newsletter, Biomedical Beat, as well as on a blog, Facebook pages, podcasts, videocasts and a Twitter feed.

NIGMS is also willing to provide a quote from an NIGMS administrator for a university's release, says NIGMS PIO Alisa Machalek: "It's good for such releases to have an extra voice that explains the significance of the research and also to have a respected institution like NIH convey their support for the work," she says.

Some private foundations also take an active role in publicizing research that they support. For example, HHMI does releases on the work of its investigators—who are employees of HHMI, but based at their home institution. HHMI releases are often quite helpful to the institution's news office, says HHMI's Jim Keeley: "My own anecdotal evidence is that many news officers are saying, 'We just did not have time to do this release; we are glad that you have one," he says. "We share our news releases, making them available to the institution, so that if they don't have their own, we are happy to provide those to them."

PIOs with Companies and Commercial Agencies

If you are at a university, and your work is funded by a company, you might find yourself working with PIOs from the company or a commercial agency they employ. Or, if you work for a small company, it might hire an outside agency or communications consultant.

In either case, understanding how to assess such firms can help you avoid the pitfalls they might present, and to communicate your work effectively and responsibly. Public relations consultant Lynne Friedmann offers advice on how to assess PIOs from an outside agency. She says the agency's writer should have a journalism background and should be a member of the National Association of Science Writers, or the Public Relations Society of America. It is also a good sign if they are an APR (Accredited Public Relations) professional, which means they have demonstrated a body of knowledge about PR and adherence to ethical conduct.

Universities also may hire commercial agencies to represent them. In working with such firms, apply the same principles of responsible communications that you would in working with your own PIO. For example, just as you would not allow your own PIO to hype or distort your work to gain attention, you should not allow such an outside agency to do so. One drawback to such commercial firms is that journalists often hold a negative perception of them.

"At the *Chronicle*, we used to get these guys coming in with fancy suits from PR agencies hired by one university or another to increase their standing," recalls Kim McDonald, who was a science writer at the *Chronicle of Higher Education* for 20 years. "The problem was that they were advertising company executives who knew very little about the substance of what was going on at the universities they represented."

Science News editor Tom Siegfried declares that "Agencies are almost always bad." The advantage of an institutional PIO, he says, is that "you have the personal individual relationship with someone who gets to know the reporters and their interests and needs personally and build a trust and helping relationship." In contrast, he says, when he has encountered agency public relations people, "it always come down to 'How do you manipulate the media into doing what you want them to do?' Very rarely has an agency ever come in and pitched something, or presented something in a way that remotely coincided with what I wanted to hear," says Siegfried. "They give us fancy press kits, and they get these weird ideas they think are clever for stories that are totally dumb as hell."

One absurd agency strategy that Siegfried encountered in his newspaper days was to offer a reporter clips from other newspapers, such as the *New York Times*, to show how important a story they are pitching is. "First of all, it's usually a story I did before the *Times*, and to think that is going to make me want to write a story because they can show you what the *New York Times* wrote, it is dumb," says Siegfried. "A message for scientists is that if they are interested in having good media relations, rule number one would be don't hire an agency," he says. "Hire a professional science journalist/PIO, who knows the field, who knows science journalism, who knows the journalists, and who knows how to cultivate those personal relationships."

You may also encounter pitfalls working with a PR company or agency on a release on your work—particularly for medical researchers with contracts from pharmaceutical companies. Says Don Gibbons,

> That first press release is something that I refer to as a ménage à trois on burlap sheets. What that means is you are excited about your work, your university PIO is excited about the work, and the drug company's PR agency is excited about the work. But they all have different reasons for being excited, and someone is going to end up with skinned knees unless you go into it carefully. You are excited about the work because of what your colleagues will think of it. Your PIO is excited because of the news peg. And the drug company is looking for something that will send their stock up. While your PIO will be looking for the caveats and making sure it's not oversold, the company will not be as interested; and you might get so taken in, you forget those caveats.

What's more, warns Gibbons, a PR agency might try to use your institution's reputation to enhance media attention to the work. For example, the agency might ask for an institution's letterhead or logo to use on a release.

Obviously, you should decline such requests. Also, make sure that the agency checks its news release quotes with your own news service, to make sure you are not being cast as an endorser of a product such as a drug.

When PIOs Synergize to Publicize

There may be occasions when multiple PIOs are working on separate releases about your work, for example if you publish a newsworthy journal article, receive a major grant, and so on. Collaborations among PIOs are often amicable and synergistic.

For example, if your PIO notifies AAAS of a planned news release on a *Science* paper, the AAAS Office of Public Programs can enhance media attention for the release, says AAAS public programs director Ginger Pinholster. The release can be posted on the AAAS online press room containing summaries of the next issue's papers, as well as on the EurekAlert! embargoed news pages.

NIGMS PIO Alisa Machalek says she considers working with a university's PIO "a win-win relationship. I have on countless occasions offered a PIO a news release to distribute themselves. I feel like we have the same goals; that we are not competing. And when we put a quote in their release, they get an authoritative quote, and we get our little piece of the story."

When a paper's authors come from several institutions, their PIOs can agree to either issue a joint release or "parallel" releases from each institution. Multiple releases or a joint release can emphasize the importance of a piece of work to journalists. At the AGU, Harvey Leifert encouraged such joint releases for just that reason.

"When a PIO would contact us about a release on a paper in, say *Geophysical Research Letters*, we would ask to see a draft in advance, with a view toward co-issuing it," he says. "And if we were doing a release, we would contact the institution to see whether they would like to come in on it. Or, if we were both doing releases, we could coordinate to issue them on the same day. Working together could multiply the number of reporters who would see the release. For example, a state university would focus on their immediate market, because they were interested in influencing state legislators; whereas, we have a worldwide distribution of about fourteen hundred reporters."

Failing to take advantage of such mutually beneficial collaborations can create strains on relationships and lose opportunities for enhanced coverage. Catherine Foster recalls when she worked at Argonne National Laboratory that there were lost opportunities to help PIOs from institutions whose scientists used its research facilities: "We always got a little upset when we saw a story go out about work done at the Advanced Photon Source, and we didn't know about it. We felt like we could have helped with it; for example posting it on our Web site. It's not that hard to drop an email to somebody that says 'Tomorrow we are putting out a news release that mentions you. Anything you could do to help would be lovely.'"

A PIO might tend to resist such coordination, "because they see it as another approval hoop, and most of us are jumping through enough of those," says Foster. "But we were not looking to approve your release, if your researcher is comfortable with it." Also, a PIO might be possessive or parochial about his or her news release, reluctant to share it with colleagues at other institutions.

In working on a news release with your PIO, your wisest course is to advocate coordinating it as widely as possible. Such coordination is to everybody's benefit, since it broadens the reach and impact of your communications.

7

Understand Embargoes, Pro and Con

Many journals impose embargoes on their papers—requiring media to hold stories on a research paper until a specific date and time. Since these embargoes affect your collaboration with PIOs and media coverage of your work, understanding them is important.

The embargo process has sparked a long-simmering controversy, as described in a 1998 *Science* article, "Embargoes: Good, Bad, or 'Necessary Evil'?" by Eliot Marshall, which describes the typical scenario of how journals manage their embargoes:

> Every Wednesday or Thursday, more than 1400 reporters around the world get a sneak preview of the research articles that will appear in *Nature* a week later. The journal sends out faxes and e-mails highlighting the most newsworthy stories, and reporters can order the full text of any article. Two days later, more than 1200 journalists get similar advance notice of articles to be published in *Science* the following week. FedEx or priority mail brings early copies of medical journals like the *New England Journal of Medicine (NEJM)* and the *Journal of the American Medical Association (JAMA)*. Reporters' e-mail inboxes and fax machines, meanwhile, fill up with announcements from other journals, universities,

and institutes promoting new scientific findings. Most of this information carries a prominent warning: EMBARGOED. Public use of the information is forbidden until a specified date and hour to coincide with a journal's publication date.

What is most remarkable about this vast private traffic in science news is that it almost never leaks prematurely to the public. Hundreds of news-hungry reporters sit on the information, as they are bidden by journal publishers, until the designated release time. Welcome to the embargo system—a gentlemen's agreement between science journals and reporters designed to manage the flow of new scientific results to the public.

Arguing in favor of this embargo system are journal editors and PIOs from research institutions and journals. They contend that embargoes give reporters time to decipher the article, interview authors and independent commentators, and prepare an accurate story. For articles on clinical advances, editors believe that embargoes give physicians a chance to read the scientific paper on an advance before patients read media stories, so they can better respond to questions.

More Machiavellian, of course, is that embargoes create some urgency and competition, which works to the advantage of journals and PIOs to spur journalists to do stories on their papers. Journalists may also use embargoes to their benefit with their editors. They may employ a looming embargo as a small stick with which to prod their editors into running a story on the "breaking" news from a scientific paper.

Critics, however, contend that embargoes skew research coverage toward articles that have an artificial news peg. In his 2006 book *Embargoed Science,* Vincent Kiernan concludes that "embargoed science appears to have a clear edge over unembargoed science, in getting into the paper and onto the air, regardless of the relative importance of the two." What's more, argues Kiernan, embargoes also warp the public's view of the nature of science: "The embargo encourages a type of science journalism that depicts research as little more than a series of isolated discoveries, with little connection to previous research and divorced from a systematic mode of investigation. The embargo, by promoting an unending stream of coverage of the 'latest' research findings, diverts journalists from covering the process of science, with all its controversies and murkiness."

Embargoes are also frankly commercial ploys by journals to promote their interests, argued critics such as Harvey Leifert, former communications director for the American Geophysical Union. "Embargoes are manipulative; they are artificial, and if you have information you should be disseminating it and not withholding it, especially not for marketing or commercial purposes," he wrote in an article in the October 2002 *Physics Today*, "Who Broke the Embargo? (It's the Wrong Question!)." He also decries the gag rule that some journals impose on scientists. By this rule, they may not discuss their research with journalists before publication, even when they are presenting that same research publicly at scientific meetings. "Scientists seem to find this normal, that a journal can tell them when they can talk about their own research, with whom and when," says Leifert. "It's mind-boggling that scientists would accept it."

The embargo process has become more complicated with the advent of rapid online publication. Although such quick publication complicates the issue of embargoes—for example, requiring more urgent production of news releases—the reality is that embargoes will be around for the foreseeable future.

Thus, your responsibility remains to understand and observe the embargo policy for the journals in which you publish. Do not let a misreading of the policy prevent you from appropriately talking to journalists. For example, says AAAS public programs director Ginger Pinholster, "We try to make it very clear that the embargo does not prohibit researchers from talking to reporters beginning the Monday prior to their publication date. Sometimes scientists, and particularly first-time authors, are so concerned with adhering to the embargo guidelines that they will rebuff reporters who call them during that time." Also, says Pinholster,

> We urge researchers to share the draft manuscript with their PIO, so that they can work together to communicate their research most effectively. And it is understood by *Science*'s editors that researchers will share manuscripts with a very small number of colleagues for initial pre-review prior to the *Science* peer review. The paper should not be distributed any further, certainly not to anyone on the Hill. And, we ask PIOs to refer reporters to *Science* to get the official final version of the paper. This gives the reporter the official version of the paper and protects the PIO and *Science* from arguments by reporters that the embargo was broken because the paper was provided by a third-party source not under the auspices of *Science* and AAAS.

Make Sure Internal Media Observe the Embargo

You and your PIO should also make sure that all the internal media understand the requirements of an embargo. For example, Machalek recalls a cautionary case in which the NIGMS was coordinating an embargoed release with a grantee institution: "A few days before the embargo date, their university magazine came out with a feature on the researcher which covered the whole story, essentially breaking the embargo," she says. "Obviously, the publication office that did the magazine was different than the news release branch, and they weren't talking to each other."

Avoid Arbitrary Embargoes

There are instances where release embargoes are not appropriate. For example, do not impose an arbitrary embargo on a news release that is not pegged to some news event, such as a published paper or report or scientific talk. For example, it is not appropriate to impose an embargo on a feature release.

And certainly do not impose an embargo on a public document such as an announcement of a news conference. National Public Radio science correspondent Joe Palca cites such an incident of what he deemed an inappropriate embargo. A PIO for a Harvard University research institute had announced a news conference in which researchers would reveal that they were proceeding with cloning human embryos using stem cells. Since Palca and his colleagues had already reported on the planned experiments, he broadcast a story saying that the institute was expected to announce at the news conference that they were proceeding with the experiments. The PIO accused Palca of breaking an embargo on the announcement that the news conference was to take place. "The PIO was telling me I couldn't talk about something that I already knew—the fact of the news conference," says Palca.

Do Post-Embargo Releases

Even if you do not have a news release ready to go before an embargo—thus missing the news peg—you should still strongly consider doing a release. For one thing, important media do not necessarily time their stories to an embargo. For example, the *New York Times* routinely publishes stories on findings long after the relevant paper's embargo is passed. And many major online news sites, such as MSNBC, readily publish stories on releases even when issued after the journal embargo.

Also, points out Machalek, many media-worthy research stories are not fodder for daily newspapers but rather for weekly or monthly specialty magazines: "I have no idea why many PIOs focus so intensely on the daily newspaper, especially for a place like NIGMS, which is less likely to be in the *New York Times* than *Chemical & Engineering News*. There are also quality monthlies out there, like *Discover*, that we are more likely to be in," she says.

Another compelling reason for doing post-embargo releases is that you are no longer restricted by the media filter in reaching key audiences. Remember you are media. Releases posted on the distribution services EurekAlert!, Newswise, or Ascribe appear on Google News, Yahoo! News, and other online news sources right along with media stories.

What's more, post-embargo releases can be picked up directly by Google, even if they are only posted on your Web site. You induce the googlebot—the software that searches out Web pages to list on Google—to directly crawl your institution's news Web pages to extract links to post-embargo releases not sent to media. To request to have your news site crawled, send a message to news-feedback@google.com. Include the URLs of the news pages you want Google to crawl. One search engine consultant suggests that sites are more likely to be crawled if they have a prominent site index.